8/22

The Agriculture Manifesto

Ten Key Drivers that Will Shape Agriculture in the Next Decade

By Robert Saik, PAg, CAC

ISBN-10: 1499382707
ISBN-13: 978-1499382709

WHAT'S INSIDE

Introduction

Red Deer, Alberta
Canada
May 2014

Technology is making a profound impact on agriculture production globally. The agriculture of the future will be very different from the agriculture of today — and vastly different from the agriculture of the past.

I've spent a lot of time assessing the key drivers that are shaping the evolution of this dynamic industry. Now, I want to get people thinking about where agriculture is going — and why it matters.

On that note, welcome to The Agriculture Manifesto. Together, we will explore and dissect the ten critical trends that are going to shape the future of agriculture — and send ripple effects across the planet — over the next decade.

This book is for a wide swath (agricultural term) of people. If you are a farmer grappling with the rate of change impacting your operation, this book will help you understand why. If you work in agribusiness and you're trying to figure out your next business move, this book will help you understand where agriculture, and indeed your business, is headed.

If you are a consumer who wants to better understand food production, this book will show you how the agriculture of today is doing an amazing job producing food in efficient and sustainable ways, and how the industry is committed to feeding our fast-growing global population into the future.

My sincere hope is that after reading this book, you'll have a greater understanding of where your food comes from, how passionate this industry is about leveraging the latest technology and innovations to continue improving productivity, and why you should continue to stay informed via credible industry sources about how farmers will continue to rise to the vital, immense challenge of feeding the world.

Robert Saik

Professional Agrologist
Certified Agriculture Consultant

The Agriculture Manifesto

Overview

Agriculture is my business, my passion, my life. It's all I have ever known. It has blessed me beyond measure. This Agriculture Manifesto is a small attempt to give something back.

I run an agricultural consulting firm (Agri-Trend) that has over 250 people who work every day with farm customers across North America as well as internationally. Through my travels, I see a lot of things happening in the world that are going to impact agriculture. Many of these trends have become strong currents that are pulling us along, often in ways we may not be aware of.

I organized my ideas into a manifesto, knowing that this would be a powerful way of sharing my vision of where agriculture is going.

In our business, we are fairly agnostic in terms of our approach to farming. We work with large farmers and small farmers. We work with conventional and very science-orientated operations, as well as organic farms. But we don't sell fertilizer, chemical, seed, equipment or buy grain. Rather, we get paid for the consulting services we provide directly to farmers at the field level. In this context, I feel I am in a great position to observe these key drivers at work.

Our model is very unique in that we have a large number of scientists and agricultural experts all rallying around the goal of helping the farmer grow the crop, sell the crop and manage the business. We do this within our fully integrated data system, so I see all parts of the equation. We touch all areas of a farm family business, whether it's agronomy and/or precision farming, grain marketing strategies, or balance sheet issues and succession planning. All of these things are going on simultaneously inside a farm business and we are blessed to have the interdisciplinary expertise to provide the necessary leadership to help farmers grow.

Plus, I'm a technology junkie! I follow what's going on in the greater society and have been tracking how that is going to pull agriculture along, very quickly. It's going to have a profound impact on agriculture production globally.

I would like to share The Agriculture Manifesto with anybody connected to agriculture, anyone who has an interest in understanding our business today, and everyone who eats food! I believe many consumers today want to have a better understanding of where their food comes from. As they read this book, I hope they may raise their eyebrows and say, "I didn't know that agriculture was so advanced and doing so many things."

This book is also meant to be thought-provoking — my hope is that it will spark a dialogue about the future of agriculture. I don't expect everybody to agree with all the points; the goal is to get them thinking about where we're going.

Ten Key Drivers that Will Shape Agriculture in the Next Decade

I'll start off with the big picture by explaining some of the experiences I've had by following people like Peter Diamandis and Ray Kurzweil as well as sharing insights gained while attending courses and traveling around the globe.

There are some overriding macro trends that are going to impact facets of society, including agriculture. If you want to understand some these significant trends, I encourage you to read the book "Abundance" by Peter Diamandis, which is one of the best books I have ever read when it comes to providing perspective on how technology and science continues to tackle the toughest challenges facing humanity, such as access to clean water, medicine, education and of course food.

Take for example, Moore's Law — the law of the microchip. Now, a $1,000 laptop is calculating at about 10 to the 9th or 10 to the 10th calculations per second. By 2023, it is possible that a $1,000 laptop will be calculating at 10 to the 16th calculations per second.

This means that a $1,000 laptop could be calculating at the same speed as a human brain!

Couple the exponential growth of computing capacity with the exponential growth of data and there will be profound changes.

It has been stated that all text from the dawn of mankind to about 2003 would be 5 exabytes of data (an exabyte is roughly 1 billion gigabytes). Today, it is estimated we are producing more than 5 exabytes of data every two weeks and soon we will be producing that volume of data every 15 minutes!

Now, combine computing power with the exponential growth of data in all fields, including agriculture. Then, tie that together with connectivity, also known as Metcalfe's Law of Connectivity, and consider that there's going to be another three billion people connected to the Internet in as little as five years. We are looking a massive change.

This means that the pace of idea exchange is going to ramp up at an exponential rate! These forces override everything we're doing in agriculture. So within that context of exponential growth, I've come up with 10 key drivers that I believe are critical to agriculture in the next decade.

I get asked all the time if agriculture can feed a population of 7 billion today and 9 billion people by 2050. My answer is always YES...if we allowed to.

So, this is where I will start The Agriculture Manifesto. The biggest threat that may prevent agriculture from eradicating hunger on the planet is the non-science movement.

1. Non-Science

Everywhere you go, whether you are in the country or the city, you're being bombarded with messages in the media about our food production, and most of these messages are negative. Most would have you believe that farms are industrial operations run by big corporations, when in fact that's not true. Ninety-seven percent of farms in North America today are operated by farming families.

Whether it's popular media, TV talk shows, self-appointed food experts and authors, health magazines or radio commercials, we are being bombarded by advertising that is designed to confuse or create fear.

People pushing their "philosophies" (and/or vitamin supplements or special diets) say you should purchase only organic and non-GMO foods. We work with organic farmers that are catering to this consumer elite and I will not argue the consumers right to spend their money where they choose. That is their choice.

But make no mistake — organic production is not sustainable for 9 billion or even 7 billion people on the planet. It's been estimated that if we go back to pre-1960's agriculture, which today would be defined as organic in nature, we would have to make a decision as to which 3 billion or so people would need to exit the planet. Science has played a key role in enabling agriculture to feed the population today, and it's being hit hard by a negative non-science or anti-science movement that's pushing a lot of negativism towards our industry.

Organic started off as a movement to reduce the utilization of synthetic fertilizers and pesticides. But to be clear — there are sanctioned pesticides allowed for use in organic food production.

Additionally, we know that plants have to eat like we have to eat. Plants have to have nutrition so within an organic food production system, those nutrients have to come from something; be it animal manure or human waste of some sort of external source. Bottom line: plants need nutrients like nitrogen to make protein and they don't care if that nitrogen comes from animal manure, nitrogen fixing plants or urea based fertilizer.

It is estimated that about half of the protein in every single human being on the planet comes from an invention that very few people understand, called the Haber-Bosch process. It is the process of taking inert nitrogen out of the air that you breathe (about 78 percent), and turning it into fixed nitrogen or what we call fertilizer. If we were to all move towards organic food consumption, we would not be able to sustain our 7 billion lives because there isn't enough net nitrogen coming into the system under an organic food production system.

I don't want to dwell too much on this but the reality is, if organic production is really what people want, then the result would be the reduction of synthetic fertilizers and pesticides. That means the next thing I'm about to say should make sense: The future of agriculture could be GMO, Genetically Modified Organic food production. But today, the GMO industry has been targeted by a very strong non-science movement. That anti-GMO movement has divided the debate between the organic and the biotechnology community.

I don't know when the organic movement became the anti-GMO movement, but I have a hunch it has something to do with the commercial interests of people and entities such as Whole Foods, Trader Joe's and Chipotle; grocery chains and restaurants that seem interested in pushing a mandate of anti-industrialization of agriculture onto consumers, which has led to a great deal of suspicion about GMO technology. In other words, I believe the non-science movement and the anti-GMO kick has a lot to do with...MONEY...and with the rise of Big Organic which grows its business on the back of fear and suspicion, not on science.

If we really wanted to be serious about increasing organic food production, then it makes sense that GMO would be a significant part of the solution.

For example, if you could breed potato plants to fight late blight and not have to use fungicides, wouldn't that be a good idea? If you could have crops biologically fix their nutrients so you didn't have to use fertilizers, wouldn't that be a good idea?

Well, GMO technology and biosynthesis promises to be able to do some of that. In fact, many of the very exciting research projects being worked on right now are in those very areas. But because of the media push, the anti-GMO movement and the non-science movement, they would rather have us drop all that technology altogether and revert backwards to some sort of a 1960's type of agriculture.

If they're successful it's going to rip away a lot of the technology that we use as farmers to grow food for society. More alarming to me — and I just spent some time in Kenya — is the non-science movement's influence on politicians worldwide.

For example, right now, millions of dollars have been spent in Kenya on a cassava crop that would be resistant to the virus that destroys it in the ground. Cassava is a staple that feeds some of the world's poorest people, and they've developed a GMO cassava that's resistant to the virus so it won't be rotten when you dig it up.

But because of political lobbying, that product sits on the shelf in Kenya — as does Golden Rice, which is bio-fortified rice with Vitamin A. This GMO rice project also sits on the shelf while millions of kids have gone blind or have painfully died due to Vitamin A deficiency that this Golden Rice could have cured.

These are examples of crops and technologies that would make a huge difference worldwide, but the anti-GMO and non-science movement is very vocal. In particular, they're lashing out and tying together corporations such as Monsanto to all GMO technology. That's true and not true, because much of the technology today has nothing to do with Monsanto or Syngenta or Bayer. A lot of the technology is being developed in public institutions or smaller private firms.

The negative spin of the non-science movement could rip the heart out of a lot of potential benefits that are required in agriculture today. I work with farmers who grow canola, soybeans and corn, and I can tell you that the technology we're implementing today at field level drastically reduces the amount of pesticides we're using to grow your food. It also significantly reduces the amount of cultivation thus increasing soil health, water percolation, infiltration and water/nutrient holding capacity.

I have spent time in the fields of Argentina and Brazil and can tell you first hand that without GMO science, their soils would be eroded below productive capacity, the world would be paying much more for food and millions more would be hungry.

Even the conversation around hormones in beef is blown out of proportion, with most facts being taken completely out of context. The hormones employed are natural, they increase feed efficiencies and their residues must be gone by harvest. The fact is that tofu is loaded with estrogen hormones at levels thousands of times higher than any beef that would be exposed to through hormone technology. The hormones you are getting naturally in your french fries are higher than those levels in the hamburger.

The last thing in this area is the spin that's being put on industrialized agriculture by companies such as Chipotle, which just did a whole series on Hulu called "Farmed and Dangerous." If you go to the Internet and search 'GMO' in Pinterest or YouTube, or in Google images, you will be hard pressed to find anything related to science. But the memes on the sites would cause you to be chilled with respect to what the non-science movement is portraying.

I think we need to fight back as an industry and be aware of what's going on around us.

It's an area surrounded by fear mongering — people are spinning fear, they're frightening parents and make no mistake, there's a very large profit motive going on behind the scenes within the non-science movement to discredit agriculture as we use it today.

The majority of these people have never been on a farm, they have never fought insects coming into a crop, they have never had to deal with weed control pressures, and they have never had to figure out how to prevent a disease from obliterating a food for the consumer.

Again, one of the reasons I'm writing this book is because that's what I do. I work with farmers at field level and the technology that we're using is very good. It's leaps and bounds ahead of the technology we used in the early part of my career in the '70s and '80s and I wouldn't want to go back. We're in a good spot.

Is it perfect? No. There are still a lot of areas for improvement but it's much better than where we were in the '80s.

The problem is; I don't see a lot of people from the science side getting the spotlight. There are a handful of individuals that I come across in social media providing a common sense approach, such as Dr. Cami Ryan, a gifted communicator, from the University of Saskatchewan. Another is Dr. Kevin Folta with the University of Florida, who is a beacon of common sense in this whole thing. The former co-founder of Greenpeace, Patrick Moore has now been challenging Greenpeace's non-science stance which is blocking the introduction of Golden Rice.

Mark Lynas, who was an anti-GMO'er among those who coined the phrase, "Frankenfood", did a complete 180 on his opinion in January 2013 and has since been an outspoken advocate on the use of GMO science and how the non-science stance has been so damaging to so many people on the planet.

There are some good sites such as GMOAnswers.com where people can ask questions and receive answers from science experts. Another good site to follow is BioFortified.org.

Groups like the Bill and Melinda Gates Foundation have also come out strongly in favor of the integration and continuing use of science and agriculture.

The reason the pro-GMO movement does not have a good voice is because science is not very sexy and many involved would rather work on doing good rather than advocating. Farmers today are working hard in the field trying to make a living and feed the world. Meanwhile, activists are spinning fear on social media and getting national attention with the likes of Dr. Oz and of course so-called experts such as "FoodBabe".

The reason they get the attention is because fear sells. People want to be able to understand; when they cannot easily understand, it is easy to choose fear. The new science is rather complex and full of jargon people can spin into a fear theme.

What people should know is that many of the crops we call "organic" today owe their existence to a widely used technique called mutagenesis. Traditional crop breeders have always used the long, painfully slow process of natural cross breeding, hybridization or natural mutation as a selection mechanism for new traits however through the process of mutagenesis new crop varieties are derived through seeds being bombarded with radiation or submersed in chemical baths to induce mutation in the plant. If the mutation is deemed to be positive, the breeders grab that change and the plant is allowed then to reproduce, and today it's a variety.

There are tremendous scientists working very hard at traditional crop breeding however due to anti-GMO pressure the use of mutagenesis is growing.

There are 2,400 varieties of organic seeds that were generated with either radiation or chemical baths. People don't know this but that's how it was done and that is how it's done still. I don't know why people would prefer that method of plant breeding to genetic engineering. People discount the fact that we have all this computing power in science today that enables plant breeding to be done much more precisely.

Why wouldn't we use that technique? It doesn't make sense to me.

We have to get better at advocating on behalf of our

industry. Each of us involved in agriculture has a responsibility to reach out and challenge the half-truths, distortions or lies being spread about our industry. In other words, we must turn ourselves into AGtivists!

2. Bioengineering

One of the huge benefits of bioengineering is the promise of better crops and more healthy food.

I believe what we need to focus on is food nutrition and nutrition density. That's a science play because there's all sorts of breeding technology that we could bring to bear that have the potential to enable plants to harness more nutrients out of the soil.

It's also an agronomic issue, because if we do a better job with our agronomics we'll be doing a better job of managing the soil, thus feeding the plant, therefore the nutritional value and the nutritional density of the crop would be better.

One of the things people will hear much more about is the basic biological building blocks A, T, G and C (Adeinine, Thymine, Guanine and Cytosine), which are the nitrogenous bases that are the keys for DNA and RNA synthesis. All around us (including in some garage-like settings) people are working with arranging these blocks to do new things in plants. This is a new area of science that holds tremendous potential while, at the same time, causing people to be fearful of the future. The point is, this science is not going to go away and its positive implementation is something we need to embrace.

There are all sorts of science coming in the biosynthesis area. GMOs would allow us to have greater insect resistance, drought tolerance and salt tolerance. All of those things are being worked on at various levels in institutions and not all of the science is GMO, as there are many other new biotechnologies also being employed.

The ability for us to understand recombinant DNA through genomic sequencing ties back to one of the big trends, which is computing technology. Today, laboratories are profiling genomes using petabytes of data to be able to isolate certain functions that are going on inside of the plants.

For example, we have the ability now to stop late blight in potatoes instead of using 6 to 17 applications of fungicide. This is a result of identifying blight resistance in a strain of wild potato then inserting that gene into commercial potatoes to give you the disease resistance. This is exciting because we could significantly reduce pesticide use and make a more profitable crop for the farmer. Sadly, it has not seen the light of day due to the non-science lobby.

Epigenetics is another exciting area. Epigenetics is not changing the biology or the genetic makeup of the parent. It's a nurture thing, in other words, we're feeding crops or feeding the livestock better, making subsequent offspring stronger. Epigenetics refers to the heritable changes expressed due to non-genetic factors.

Basically, if you take two offspring and feed one of the offspring well and starve one of the offspring, the children of that well-fed offspring will express themselves more strongly because the parent was well fed. For example, we are learning plants that are grown on high zinc soils will not only have more zinc in the crop, but will also be stronger when subsequently seeded.

Biocatalysts, hormone technology, polymers and stimulants are all helping us increase the plant's ability to fight abiotic and biotic stresses. I see more and more technology stacking coming into play where we take bits of technology from a variety of areas and stack them together — whether they be agronomic or precision farming, computing or fertility. All of these things stack together allowing us to do more.

Biofuel is another area with tremendous potential. Keep in mind that most of the biofuel technologies we read about are first or second generation. I think there'll be a lot of tertiary technologies coming down the pipe. I'm excited about the potential of algae.

Creating fertilizer from waste is another area that is going to continue to grow, and as a result, we will be able to make better use of our biological wastes, whether they're urban or rural.

I'm very excited about other areas of plant science, such as proteomics, which is the large scale study of proteins in plants, or ionomics, which is the study of ion manipulation in leaves helping to determine how plants respond to stresses. These are technologies that people are working on every day.

The ability of us to feed the planet will depend on how we battle abiotic and biotic stress in crops and livestock. As exemplified by ring virus in papaya, citrus greening in oranges or potato blight, Mother Nature never stops challenging our food production systems. Without embracing the science of biosynthesis, we will be hard-pressed to meet the needs of humanity. But with the new frontiers and promises of bio-tech the future is bright indeed.

3. Market Segmentation

Contrary to what you might read, there are still significant areas to expand actual arable land on the planet, but the reality is that we have to do more with what we've got, which leads us to market segmentation and niches.

I believe that when we're looking at agriculture as a whole, we need to think about what we're growing, who wants to eat that food and what they can afford to pay for it.

We can start with the 'locavore' movement. The locavore movement, exemplified by the increased popularity of farmer's markets, is very strong in many areas and has an important role.

Most people in urban communities today are disconnected from farming operations. They want to be able to connect with the food. This local movement, and the organic movement, is very strong in many areas. It's a niche, and one that can be filled by farmers.

I believe, however, that some of the larger opportunities lie in focusing in on city markets. Countries such as Canada and the United States export their agricultural commodities to other countries. Well, I'm looking around the globe thinking, rather than trying to crack South Korea, maybe we just try to focus on Seoul.

Or instead of trying to crack China as a market, maybe we just need to focus in on one city in China. Instead of trying to focus a specific crop that I'm growing and trying to sell it into the US market, maybe I need to focus in on Boston or New York City. One of the marketing campaigns that is gaining ground is niche or targeted markets.

This is being driven by our ability to track how we're growing the crop and being able to substantiate it. Companies such as Walmart, Sysco, Costco and McDonald's are increasingly concerned about sustainability. Most farmers roll their eyes when they hear that term because they've always been practicing sustainability. But the reality is that consumers want us to <u>prove</u> sustainability.

A majority of these companies are publicly traded. They're going to come under the scrutiny of the public and they want to be doing the right thing. Again, this is not known to most people but there are sustainability initiatives all over the United States and Canada that deal with those issues I mentioned before such as efficient water use and soil quality.

We've been working on helping farmers understand the attributes of sustainability, such as soil quality, carbon footprint and water utilization. We then help them measure and document this key information so we can show the consumer, or show the food processor buying our commodity, that it is grown in a sustainable manner. I believe that creates one of the new opportunities for us going forward.

We have to differentiate between local and commercial markets. This is why I find this non-science movement so unpalatable. If you choose to support a non-GMO or an organic food, that's your business. But don't tell the rest that it's bad because if you obliterate that science, we won't be able to feed the rest of the population that goes to the grocery store and buys food.

I think the "local" versus the "commercial" markets are different. Small farmers of two or three acres meeting a local need must charge high prices and are vastly different than the commercial operations, which are farming thousands of acres and producing crops and animals in a more commercial manner.

We need to understand that there has to be different messaging for different markets but the messaging should not build one sector by tearing down another.

In North America, people who want to grow their own food or buy from local producers can do so with the confidence of knowing they are back-stopped from hunger by the commercial family farmers who supply the Costco's, Safeway's, etc.

Another exciting trend I see happening is growth of trait-based technologies. In other words, to be able to target what we're growing for the demands of a specific market. For example, if we could grow wheat that was high in selenium and you could prove this, then you could you make that into bread that you would sell at a premium to men because selenium is known to help mitigate prostate cancer.

This is a good example of science dovetailing with trait tracking and getting nutrient density to a specific market. These are the things that farmers and companies need to think about when marketing their agricultural products.

I also see opportunities for farmers to do direct shipment from the farm to the end consumer. As an example, Saskatchewan is one of the largest producers of pulse crops in the world, specifically chickpeas and lentils. Many farmers today could take a Sea-Can container fill that up with red lentils, lock that container and send it all the way across the ocean or across the world to India or Pakistan where it would go directly to a consumer.

Stick a QR code onto the container that ties everything about how that crop was grown and you have full traceability.

I see more of that happening because of the tracking ability, and our ability to meet with and connect with buyers on the other end of the world who are after specific attributes. Containerization and miniaturization of the shipments is something we're going to see more of going forward.

Nobody, Monsanto included, wants to put the crop production of the planet at risk. Nobody wants to have a food supply that would result in detrimental damage to the population. Why would any farmer get up in the morning and put something on the crop that he wouldn't feed to his own family? It doesn't make any sense.

I think the larger corporations are very concerned about this. They have to be thinking about being able to feed the broader population so they need the quantity of food. Yet, they want to be able to hold their heads high and know that they could stand up to public scrutiny if people wanted to know how that product was grown.

4. Sensor technology

The next section of The Agriculture Manifesto is about sensor technology.

I see sensor technology absolutely exploding in agriculture. This includes in-field sensors such as soil moisture probes that we bury in the ground to determine whether soil is at field holding capacity or at wilting point. This technology can 'ping' the farmer on his handheld device to let him know that his field is drying out. Then, he could consider turning on his irrigation or even having that irrigation system autonomously turned on.

We are currently integrating grain bin tracking in our data systems that can be tied to sensor technology that will alert a farmer when grain temperatures or moisture levels are climbing to a level that might lead to spoilage of the crop. This bin sensor technology can also alert a farmer when the grain is being removed from the bin and might help reduce on-farm theft.

Other technologies coming down the pipeline are nitrogen and nutrient sensors that we can bury in the soil which allows farmers to watch what's happening and better understand what the potential of the field is today. I was reading about a brand new technology that basically sniffs the air and does a DNA analysis of the air to find out if there are spores of disease blowing onto the fields!

Sensors would alert us to what kind of protective measures we could take to help the crop fight the disease before we would ever see the disease on the plant.

These kinds of sensing technologies are everywhere. We are working with satellite technologies that allow us to monitor the field conditions from the sky. In particular, micro satellites will have a huge impact on agriculture. These are low-orbit satellites providing us with faster refresh rates than the big satellites, so we can get a picture of a crop while it's growing in real time. We would be able to compare the crop on July 1, July 5 and July 10 to find out if there are any abnormalities or differences in that crop.

By building algorithms, we may be able to alert farmers when there's a change happening in that crop. This would allow farmers and their consultants to go in and investigate these issues. We can do this by walking into the crop, or we could simply send a UAV (Unmanned Aerial Vehicle) or drone over the crops. Drone technology is something we're playing with right now, which of course is better than satellites because satellites are encumbered by cloud cover, whereas drones can be put in the air to scout a field any time. That drone technology allows us to be able to isolate or localize where the infestation and problem is happening in the field.

Then, just think about where robotics is going. We might be able to have robots actually going into the field and doing this sampling or some of the monitoring for us. Ultimately, these sensor technologies are going to be everywhere in agriculture. We're seeing them in equipment right now where the equipment is sending a signal to the dealership letting them know that the bearing is getting hot and the dealership should send out a repair truck to replace the bearing before it ceases to function.

These technologies are rampant in agriculture today. Anything could be sensed — from oxygen and carbon dioxide levels inside of greenhouses, to water levels, nutrient levels and even disease levels in crop.

This is a brand new field that is going to explode in the next decade.

5. 3D Printing

The next trend I want to talk about is 3D printing, which is going to be very interesting.

If you followed the development of 3D printing, when it started if was very expensive. Today you can buy a 3D printer for $1,500 to $2,500 and it will produce 3D items for you in your home office.

But increasingly, what we're seeing is the printing of biosynthetic parts. If you check out 3D printing on YouTube you will see researchers right now 3D printing ears and noses.

The basic structure of 3D ears and noses are built with some plasma from your own body. You give the scientist a sample and they use that as part of the substrate to build the architectural framework for a new ear or a new nose. This gets around one of the major problems in medicine, which is rejection.

You're going to see the replication of parts continue, but if you also flash forward, would you start thinking about how we could biosynthesize food. In other words, could we biosynthesize a steak? If you're averse to the harvest of animals to produce your juicy T-bone, could we instead, biosynthesize that meat?

In fact, right now they've already biosynthesized the first hamburger — and apparently it tastes like hamburger. It only costs $300,000 but that cost will come down over time. In the future, this 3D printing is going to very much emulate a Star Trek Replicator. One of the opportunities is to be able to regenerate parts in real time.

If you think about this from a farmer's standpoint, imagine that you're combining and suddenly a pulley breaks. You've got to stop the combine and go into the equipment dealership in town, find out if he's got the part. If he doesn't have the part, he has got to go to the regional warehouse or he's got to go to the manufacturer, and you could be down three or four days before you get that pulley.

In the future, 3D printing is going to allow the company to send you the file and maybe that 3D pulley will be printed out right in the farmer's shop or at the dealership in town. Instead of stalking all of these parts and trying to figure out what's going to break, all you do is have the raw material at the site and the raw material 3D prints out the part.

I think this is going to revolutionize the parts and the equipment industry going forward — and it's going to drastically change how we handle our inventories and the delivery of repair parts to farmers.

It's inevitable that we're going to move in that direction. Sooner or later someone is going to 3D print kidneys and 3D print livers and if you can 3D print a kidney or a liver for a human being, how long will it be before you're 3D printing meat for consumption?

That begs the question: What are the ethics around that? Would you embrace it or not? It's really strange. You watch the consumer react when I ask them, "Would you accept a 3D printed steak?" Their faces scrunch up and they say, "No, I don't want that. I want real steak."

But if you couldn't tell the difference, what would that mean to the livestock sector? That's one of the things I'm thinking about.

6. Robotics

Google has at least ten driverless cars that have put on over half a million kilometers in places like Los Angeles and Las Vegas without an accident. We have GPS guidance technology at farm level today which is sub-inch accuracy. That means we have the ability to take an 80-foot planter and guide that planter, using GPS technology pulled by a tractor, to within an inch as it is going up and down the field. So one would have to ask the question: How long is it going to be before we have unmanned tractors?

I would predict that it's going to be sooner rather than later. One of the greatest problems we have today in agriculture is labor, finding qualified people to operate these half-a-million-dollar pieces of equipment.

Right now the sprayers that we use in the field are up to 120 feet wide. These sprayers are traveling up to 18 miles an hour, applying crop protection products over the field. The reason that the equipment is so wide is that we're having difficulty finding qualified operators to run these large, complex machines.

Instead of having 120-foot wide sprayers, maybe in the future we'll have 120, 1-foot sprayers that are traveling up and down the field in a swarm, controlled autonomously by robotics that will do the work for us. I believe that eventually we'll hand over a lot of the redundant work in agriculture to robotics.

Another area where that's really prevalent is the milking parlors. Increasingly, robotics are integrated into milking parlors to take over the physical labor associated with milking cows. The early attempts were cumbersome but today that technology is widely adapted throughout the world.

Another interesting thing about robotics and automation is the ability for the equipment to tell us what's going on in the field. As a professional agrologist, one of the most difficult things I do is to find out what was actually applied to the field. I'm counting on the farm operator to tell me what was applied, but there are a lot of variables to that information getting back to me.

What if we were using wireless data transmission? What if the machine was automatically transmitting the data of what was applied in the field to the cloud, and subsequently into our database? I can tell you that this is actually happening today. The data system we are using in our business (Agri-Data) has been linked with John Deere combines for full wireless data transfer so that when a combine is going across the field harvesting the crop, we're getting the data sent wirelessly from the combine into our database through a linkage with myjohndeere.com.

That changes everything, because it allows us to have an unmanned tractor that is running over the field, and the data — concerning the products or the seeding or whatever operation is being performed — is being transferred back into our computer and data system in a wireless data transmission. That's going on today, which is just amazing.

You take all of that integrated with artificial intelligence, which would enable the equipment to start making decisions about what it should do and those kinds of things are all going to be possible right from the field.

While robotics will help, farms will never be-unmanned because they are a biological system that requires continuous tweaking but we are moving in that direction that stretches the expertise we do have.

Less than two percent of the population in North America actually works on a farm. It's getting harder and harder for farmers to find qualified people. The only way I could see out of this whole thing is what I'm seeing in other parts of the world, which is automation of a lot of the processes that are going on at the farm level.

This trend is already happening in many sectors today. Key aspects of our operations have got to be automated, and robotics is going to play a huge role in that automation.

One of the promising areas is harvest. With human operators if, even the weather is good but you've already put 16 hours in the seat of the combine, you may want to do another four or five hours but you are exhausted.

You either push yourself and have accidents, or you stop the combine. Well, robots don't care. A robot is going to get up at 3 am if the conditions are right, and it will keep working until 3 am the next morning. There are lots of things that are going to make this a reality.

Already in Europe we have Fendt tractors with 'a master and a slave,' or essentially two tractors. The second one does not have a driver in it; it just follows the first tractor. John Deere has developed what is essentially a Wi-Fi system around the combine. As the tractor and the grain cart pull up to the combine, the driver releases control of the tractor.

The combine then takes over; the tractor and the grain cart become a slave to the combine while it unloads. The combine signals the grain cart how to go forward and backward in order to fill up. This is going on right now.

The biggest advantage to the integration of robotics on farm will be in the areas of labor savings, safety and productivity. It will enable those who really know what needs to be done, to do more because they will not have to do it.

7. Water

The next area of opportunity is water. Agriculture uses over 70 percent of the fresh water on the face of the planet today, so it's absolutely incumbent upon us to utilize technologies that will do a better job of water use efficiency.

Water use efficiency is everything. If I had to pick one metric to measure agricultural production worldwide, it would use water use efficiency. To grow the feed and supply the chicken, it takes about 120 gallons of water to make one egg. One watermelon takes about 100 gallons of water. One pound of meat takes about 2,500 gallons of water.

Everything that we have in agriculture is a function of water. Our ability to utilize science and technology to better meter out and use water is absolutely critical. Things such as biosynthesis of crops that could make better use of water by reducing evaporation rates from the plant itself. It could also help scavenge more water from the soil.

We can also get a better understanding of what the soil moisture levels are with moisture sensors, allowing producers to improve irrigation efficiency, reduce water loss, or get a notification indicator that there's a drought coming. We can now tell if we are losing water holding capacity in the soil, indicating that a farmer might want to apply some plant hormone technology that would drive roots out, enabling the plant to have a chance to withstand another five or ten days of drought or turn on irrigation.

As we coach farmers we work at educating them on how balanced fertility programs are important in water use efficiency. Potassium and to a lesser extend chloride and boron all play a role in reducing transpiration losses. By properly balancing out the fertilizer using soil tests and sensor technology, we can help the plant "sweat less". This means we can grow more corn, carrots or canola from an inch of given moisture thus increasing the water use efficiency.

The key to sustainability in agriculture is water, and everything that we can do that would allow us to utilize water more efficiently for agriculture is a step in the right direction.

I believe this is one of the key metrics we need to measure globally in order to provide a consistent measure of sustainability in our industry.

8. Precision Agriculture

Precision agriculture is all about being able to put the right amount of product at the right place at the right time in the right source. Whether it would be a variable rate seed, variable rate fertilizer, variable rate crop protection products or variable rate irrigation, we are going to see this whole concept of variable rate technology or precision farming explode in adoption.

Our work in variable rate fertilizer has shown the average net benefit to be somewhere between $35 and $45 per acre. This is the average productivity or profit gain to the farmer, versus applying a straight fertilizer rate across the field.

These Canadian and US results have been echoed by others in the field, including one of our colleagues, Moe Russell out of Des Moines, Iowa, whose studies arrived at similar results in the United States. Without a doubt, one of the major trends in agriculture is the movement towards precision agriculture.

Precision ag could also include varying the amounts or variety of seeds from the hilltop to the lowest part of the valley, so as we move through these microclimates we're going to be making decisions on the types of seeds that should be put in that place. There are many companies working towards this, so we are gaining a greater understanding of what different varieties do according to the microclimate. Many are working on collecting weather information for different parts of the field in order to understand what's happening in these different zones.

We're using satellite technology right now to do historical assessments of crop productivity indexes to help farmers establish long term production zones. We're also soil testing those various zones and then custom prescribing the nutrients that are required for different parts of the field depending on the yield that the farmer is trying to attain in each area.

Shoving those recommendations into a rate controller that's guided by GPS technology in the field allows the planter automatically change the amount of fertilizer being applied, depending on location and depending on the target yield is for that particular area. This is resulting in a net savings of fertilizer or better utilization of our nutrition. It's saving the farmer money while putting our resources in the proper place.

This technology is going to continue to ramp up. In fact, if you harmonize this technology with remote sensing such as micro satellites, it opens up the farmer's ability to do things like variable rate fungicide, variable rate crop protection and of course, variable rate irrigation, which increases water use efficiency. As we know, not all parts of the field need the same amount of water because soil structure changes and evaporation rates inside the field change.

This is all tied together with trends such as remote sensing and wireless data transmission, which allows us to know what's going on in the field. Precision agriculture allows us to do everything on the farm more precisely and this is a movement that's going to continue to grow and cause separation at the farm level.

Farmers who adopt this technology are going to move ahead at a more rapid rate. Ultimately, we're going to see a gulf between these progressive producers and the farmers that are resistant to adopting technology. It could create quite a chasm in agriculture.

9. Artificial Intelligence

Artificial intelligence has to do with the improved decision-making capability we'll have as a result of the exponential amounts of data being collected.

The question with all this data is; how will we make the decisions? Initially this is going to create a lot of confusion in the marketplace because we are gathering a lot more information than we can make decisions with. For example, we have been working with our farm customers to track attributes on million acres of farmland. We don't even know what to do with all the information we've gathered, but we know that at some point we'll be able to have help in terms of decision making, and this is what artificial intelligence is all about.

Artificial intelligence will assist us in sifting through and dissecting this data so we can use it to help us make decisions. Algorithms, which we hear about all the time, will become increasingly a part of our lives because without algorithms, we won't be able to pick out the anomalies or make decisions inside of the data.

Even today we are building "intelligence" into our data analysis in order to help make better recommendations for farmers. There are so many variables to consider in crop production; even the area of making a fertilizer recommendation has hundreds if not thousands of variables that ultimately dictate how a farmer should fertilize his crop.

Eventually, we are going to have some types of interfaces being built between human neurons connected to digital circuitry that will enable enhanced intelligence or more rapid decision making.

I think that soon, human beings will be able to tell a computer what they think about a fertilizer recommendation, seed selection or pest infestation and have these ideas cross referenced by a virtual agricultural expert. This will go a long way towards developing autonomous recommendations linked to equipment so this technology will essentially be able to take care of these problems in the field all driven by sensor data.

We're also going to have more augmented devices in agriculture. For example, you could go into a field and put on Google Glass, look up at the sky to see what the weather is like now — or in the next 24 or 48 hours. Then, you could look down at the soil and be able to get a soil test, historic soil test levels and maybe moisture levels. Next, you might look at the crop and instantly know the day the crop was planted, the crop population and many other types of information.

Being able to look at the leaf of the crop and see the tissue test levels of the nutrient status inside the leaf — all instantaneously inside of Google Glass — will help farmers make decisions. I believe we will see a big increase in augmented reality devices on the farm. They will be able to collect data, and also help us interpret the data around what's happening in the physical world.

That is going to be very important because as we gather exabytes of data about agriculture, we will need artificial intelligence to help us make better decisions on how to conserve water, use nutrients or increase the amount we're producing with a given amount of resources.

10. Data

The last trend is data and this is the biggest opportunity in agriculture. The reason I'm so excited about the capture of data is because data is growing at an exponential rate.

However, that data right now often is very disparate, in other words it's dispersed in a variety of databases.

From an agricultural perspective, it's going to be exciting times! It will also be an era that is fraught with some angst and questioning about how we should structure this.

I'm a big believer that farmers should own and control their data — that the data systems they're employing on the farm should be "farmer-centric". I have no doubt that large corporations need access to the data to verify their products and figure out how certain varieties perform under certain enviro-climatic conditions. But I also think that the data needs to be farmer-centric. I believe that congruency of the data is key.

Data exists right now on farms in Excel spreadsheets, in coiled notebooks, in Word documents and a variety of accounting programs. I believe this data needs to exist in one integrated set and that data needs to be able to move seamlessly between growing the crop, selling the crop and managing the farm business.

I don't believe there should be separate data systems to handle all of this. Farmers need to look for congruency or connection in their data. An integrated strategy is really important.

The data that we collect is going to change all of our decisions because if you can't measure it, you can't manage it — and the management and measurement of the data is what will allow farmers to make better decisions on how we produce food for the world. The data that we collect is going to allow us to grow crops and raise livestock better. Initially, we may not know what to do with all of the data but eventually, trends will emerge and advancements will be made, all contributing to better decisions.

If society wants farmers to feed the planet then we have to be able to measure what's going on in the planet. When someone says to me that developing nation farmers won't be able to utilize this kind of technology, I say they're wrong because in the next five years or so, another three billion people are going to get smartphones. I was traveling in Kenya where a family would struggle to get water for the day but they had a cell phone in their hands! They're able to text and communicate, so the measurement of things with these devices is only going to go up. This is why I think the data is going to be critical to agriculture going forward.

There is a new breed of farmer out there grappling g with all these technologies.   It's changing the dynamics on the farm.

To illustrate: "If the combine just quits during harvest, who is the 45-year-old farmer going to call to get it going, his 72-year-old father or his 17-year-old son?"

This is the world of a 5000-acre Western Canadian canola farm, or a 3000-acre corn and soybean farmer in Iowa, or a cotton grower down in Mississippi. This is the world of a large vegetable producer in California.

This is the world of agriculture and many people in society today have a very limited understanding or appreciation for the amount of technology that we're using in agriculture.

What I'm sharing with you is a glimpse into the future — but it also is very much rooted in the present; it's what is happening right now.

How to Get Involved With the Future of Agriculture

I hope you have enjoyed The Agriculture Manifesto. The idea of this short book was to inform and perhaps spur you to want to learn or do more.

If you question what I am saying...good!

If you don't believe me, go online and start to source out the facts. Or, better yet, talk to a real, live farmer and gain some insight as to what is really going on at field level.

Try to stay away from some of the rhetoric and wild stuff out there that keeps being repeated. Instead, go after the science and you will find out what I said is true.

I am also available to speak at industry or consumer events about the science of agriculture and The Agriculture Manifesto.

In addition, you could sign up for a free subscription to a magazine we publish called the AgAdvance Journal, or visit agadvance.com. The Journal is an agricultural cross between Harvard Business Review and Popular Science.

From a producer standpoint, I'd like farmers to consider working with Agri-Trend and our coaching network. We've got a vast number of coaches in all different disciplines that are helping farmers grow the crop, sell the crop and manage the business.

Then from a data management side, there is our Agri-Data Solution Platform. Agri-Data Solution is one of the most robust data platforms in all of agriculture. Farmers can use this data system to manage all of their farm production data. It connects with all sorts of outside sources, sensors and equipment — but is farmer-centric, in other words the farmer is the one who is in control of the data.

Thanks for joining me on my journey!

Author Bio

Robert Saik, CEO of The Agri-Trend Group of Companies is a Professional Agrologist and a Certified Agricultural Consultant. As founder of The Agri-Trend Group, Robert has been involved in the development of many new agricultural business processes and concepts. Agri-Trend has been nominated one of Canada's 2012 Top 50 Best Managed Companies and was recognized by Venture Magazine as one of Alberta's 2013 top 25 Most Innovative Organizations.

His agricultural travels have taken him throughout North America, Brazil, Argentina, Russia, Ukraine, Kazakhstan, Kenya, Europe, New Zealand, Australia, Chile and Peru.

Robert is a Director of Westerner Park, 2014-2015 Chairman of Agri-Trade Show and serves on The Red Deer Chamber of Commerce Ag Policy Committee as well as Advisor to The Canadian Management Council, The Farm Progress Show and The Red Deer College Donald School of Business.

In 2006 he was recognized as Distinguished Agrologist of the Year by the Alberta Institute of Agrology. He is passionate about pursuing business opportunities in the Agricultural Sector.

How to Stay Informed About the Future of Agriculture...

Whether you are;

- a farmer who wants to stay connected with consumer trends
- an agribusiness person interested in where our industry is headed
- or a consumer trying to separate hype from truth.

I am willing to share what I see out in the field every day. The good news is, I am profoundly optimistic about the ability of agriculture to feed our planet, despite the population growth, because I see what science and technology are doing to improve our food production and because every day I work with dedicated, passionate people who care deeply about agriculture, the world's most important industry.

Here are Five Ways You Can Get Connected and Stay Informed About What's Really Going On in the Agricultural Industry Today and Tomorrow...

Option 1: Sign up for a free subscription to AgAdvance™ Journal – AgAdvance.com
Option 2: Work with an AGRI-TREND® Coach – AGRI-TREND.com
Option 3: Become an AGRI-TREDND® Coach – JoinAGRI-TREND.com
Option 4: Control your data with The Agri-Data® Solution Platform – AGRI-DATA.net
Option 5: Have Robert Saik be a keynote speaker at your event – Rsaik@AGRI-TREND.com

Most people believe the hype they see on the 5 o'clock news. Fear sells. It's easy to confuse people when you only have fragmented information.

Now you can get the facts and separate them from fiction.

If you'd like us to help, just send an email to: Trends@AGRI-TREND.com

I have a blast on Twitter and hope you follow me @RSaik

58157578R00031

Made in the USA
Lexington, KY
04 December 2016